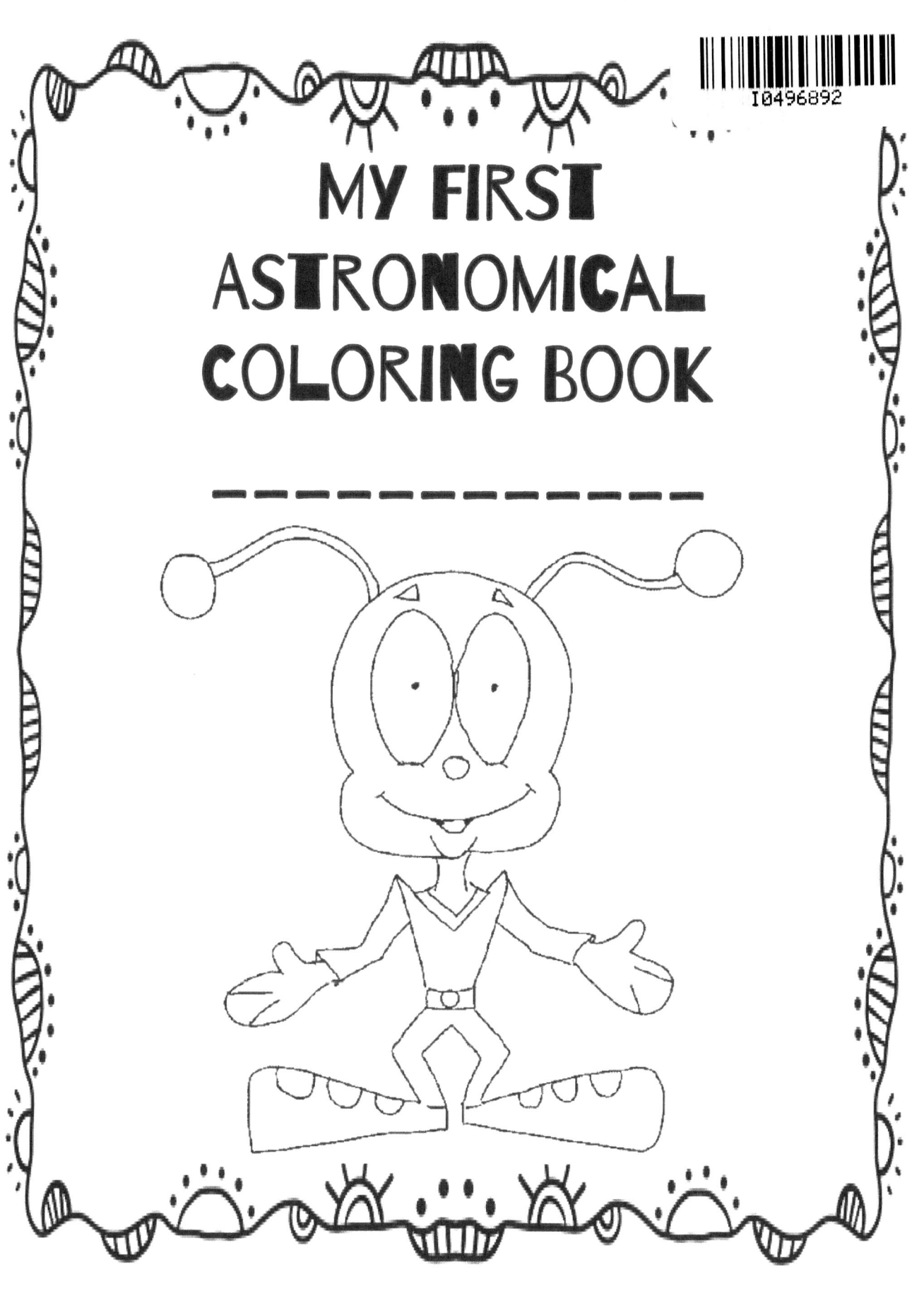

This page is intentionally left blank to for bleed through.

This page is intentionally left blank to for bleed through.

This page is intentionally left blank to for bleed through.

This page is intentionally left blank to for bleed through.

This page is intentionally left blank to for bleed through.

This page is intentionally left blank to for bleed through.

This page is intentionally left blank to for bleed through.

This page is intentionally left blank to for bleed through.

This page is intentionally left blank to for bleed through.

This page is intentionally left blank to for bleed through.

This page is intentionally left blank to for bleed through.

This page is intentionally left blank to for bleed through.

This page is intentionally left blank to for bleed through.

This page is intentionally left blank to for bleed through.

This page is intentionally left blank to for bleed through.

This page is intentionally left blank to for bleed through.

This page is intentionally left blank to for bleed through.

This page is intentionally left blank to for bleed through.

This page is intentionally left blank to for bleed through.

This page is intentionally left blank to for bleed through.

This page is intentionally left blank to for bleed through.

This page is intentionally left blank to for bleed through.

This page is intentionally left blank to for bleed through.

This page is intentionally left blank to for bleed through.

This page is intentionally left blank to for bleed through.

This page is intentionally left blank to for bleed through.

This page is intentionally left blank to for bleed through.

This page is intentionally left blank to for bleed through.

This page is intentionally left blank to for bleed through.

This page is intentionally left blank to for bleed through.

www.ingramcontent.com/pod-product-compliance
Lightning Source LLC
Chambersburg PA
CBHW060441220526
45465CB00008B/3229